QUANTUM GRAVITY IN A NUTSHELL2:

BEYOND HAWKING

The Cosmic Quest for the Quantum Theory of Gravity

Balungi Francis

Copyright © 2018 by Balungi Francis. All Right Reserved. No part of this publication may be reproduced, distributed, or transmitted in any form or by any means, including photocopying, recording, or other electronic or mechanical methods, or by any information storage and retrieval system without the prior written permission of the publisher, except in the case of very brief quotations embodied in critical reviews and certain other noncommercial uses permitted by copyright law.

Visionary School of Quantum Gravity
Independent Publishers
Kampala, Uganda, East Africa
+256 (0) 777 105 605
Email: bfrancis@cedat.mak.ac.ug, balungif@gmail.com
Second Book Edition: 2018. With new chapters

Contents
Dedication   4
Introduction   5
New Physics   9
Dark Matter and the Cosmological constant Problem   17
Why does the Zero-Point Energy Of the Vacuum not cause a Large Cosmological Constant? What Cancels it out?   23
Proof of Newton's Law of Universal Gravitation   26
Revised Gravitation Theory for the Modified Newtonian Dynamics (MOND) Paradigm and Beyond   28
Planck Stars   33
References   40

Dedication

To Abba Father for Guiding me towards the End of this Book.

Introduction

History tells us that if we hit upon some obstacle, even if it looks like a pure formality or just a technical complication, it should be carefully scrutinized. Nature might be telling us something, and we should find out what it is (G. t Hooft, 1997).

In physics, one of the ultimate goals is to unify the fundamental forces of nature. Today physicists have been able to unify three of the four known fundamental forces. The electromagnetic, the strong and the weak nuclear forces are described in a single quantum field theory, the standard model. The fourth fundamental force, gravity, on the other hand is described by the general theory of relativity. Since the other fundamental interactions are quantized, it therefore seems natural that in a grand unified theory, a theory of all the fundamental forces, gravity is quantized as well into perhaps Quantum gravity.

A theory of quantum gravity is needed to describe things that are very small but also very heavy, like black holes or the early universe. Yet gravity is described by Albert Einstein's theory of general relativity , which seems fundamentally incompatible with quantum mechanics. This is because in general relativity all physical qualities have definite values, whereas in quantum mechanics they do not. This is shown in Heisenberg's uncertainty principle.

General relativity predicts two kinds of singularities; the cosmological singularity at the beginning of our universe and the singularities at the centre of black holes. However the appearance of singularities in any physical theory is an indication that either something is wrong or we need to reformulate the theory itself. Singularities are like dividing something by zero. The problems in GR arise from trying to deal with a universe that is zero in size (infinite densities). However, quantum mechanics suggests that there may be no such thing in nature as a point in space-time, implying that space-time is always smeared out, occupying some minimum region. The minimum smeared-out volume of space-time is a profound property in any quantized theory of gravity and such an outcome lies in a widespread expectation that singularities will be resolved in a quantum theory of gravity.

However, Prof Brian Dolan at the Department of Theoretical Physics, NUI Maynooth, is quick to point out that there is not yet any set agreement on what a theory of quantum gravity should look like, or even on the exact problem it is trying to solve.

"There is no accepted theory of quantum gravity," he says. "There are currently a number of contenders, and by far the most popular is superstring theory. Many physicists find superstring theory compelling due to its internal elegance, but despite decades of intense research it has not produced a single experimentally testable result."

He suspects that trying to unite general relativity and quantum mechanics may be the wrong way to go, and that any future breakthrough may come from a completely unexpected direction; perhaps from some young mind with a fresh perspective.

"We are probably asking the wrong questions at the moment," he says. "Nevertheless it is impossible to resist the temptation to try. After all, the other fundamental forces – except gravity – fit very neatly with quantum mechanics."

Beyond Hawking is a book in a series of quantum gravity in a nutshell that employees new ideas towards the development of a quantum theory of gravity in a bid to solve the following unsolved problems in physics;

a) Is it true that at every spatial dimension there exists new physics and that it is the work of Physicists to find out? What is the method or procedure through which new physics can be found? Does this require extra dimensions?

b) Does nature have more than four space-time dimensions? If so, what is their size? Are dimensions a fundamental property of the universe or an emergent result of other physical laws? Can we experimentally observe evidence of higher spatial dimensions?

c) What is the cause of the observed accelerated expansion of the universe? Why is the energy density of the dark energy component of the same magnitude as the density of matter at present when the two evolve quite differently over time; could it be simply that we are observing at exactly the right time? Is dark energy a pure cosmological constant or are models of quintessence such as phantom energy applicable?

d) Can the singularities that plague the General theory of Relativity be resolved in any quantum theory of Gravity?

This wonderful and exciting book is optimal for physics graduate students and researchers.

New Physics

Regularization and Physics Beyond the Standard Model

The Standard Model is inconsistent with that of general relativity, to the point that one or both theories break down under certain conditions (for example within known spacetime singularities like the Big Bang and the centers of black holes beyond the event horizon).

The appearance of singularities in any physical theory is an indication that something is wrong and that there is a need for new physics. Singularities can be avoided in GR and any field theory through the introduction of an efficient regularization procedure as this book directs.

Regularization is a method of modifying observables which have singularities in order to make them finite by the introduction of a suitable parameter called regulator. The regulator, also known as a "cutoff", models our lack of knowledge about physics at unobserved scales (e.g. scales of small size or large energy levels). **It compensates for the possibility that "new physics" (beyond the SM) may be discovered at those scales which the present theory is unable to model,** while enabling the current theory to give accurate predictions as an "effective theory" within its intended scale of use.

The need for regularization terms in any quantum field theory of quantum gravity is a major motivation for Physics beyond the standard model. Infinities of the non-gravitational forces in QFT can be controlled via renormalization only but additional regularization and hence new physics is required uniquely for gravity. The regularizers model, and work around, the breakdown of QFT at small scales and thus show clearly the need for some other theory to come into play beyond QFT at these scales. A. Zee (Quantum Field Theory in a Nutshell, 2003) considers this to be a benefit of the regularization framework, theories can work well in their intended domains but also contain information about their own limitations and point clearly to where new physics is needed.

Therefore the main objective of this section is to discover new physics at those scales (or extra dimensions) which the General relativity theory and Quantum mechanics is unable to model. The section also sets out to prove that due to quantum gravity effects, there must be a minimum distance beyond which the force of gravity no longer continues to increase (operate) as the distance between the masses become shorter.

General Theory

During the years, strong evidence has appeared that the acceleration of any physical object cannot be arbitrarily large, but it should be superiorly limited. For example in string theory, it was derived that string acceleration must be less than some critical value, determined by the string tension and its mass. From the classical point of view (as

Wheeler suggested), if we consider an extended object in **rotating motion**, we have the acceleration $a = v^2/R$ and it follows that a, must be at least limited by c^2/R. However to differ from the classical Newtonian mechanics and Einstein's General relativity theory we introduce a regulator "Cutoff" $\alpha_g{}^n$, where α_g is the gravitational coupling constant, R is the distance between two masses and n is a positive number (**extra dimension** number), then the acceleration must be limited by $a = \frac{c^2}{R}\alpha_g{}^n$ (i). In section 3 we show evidence for maximal acceleration and singularity resolution in Quantum Gravity for both the Reissner Nordstrom and Schwarzschild radius.

*Thus to avoid the infinity but while retaining the point nature of the particle would be to postulate a small additional dimension **n** over which the particle could 'spread out' rather than over 3D space.*

The idea of including extra dimensions, to achieve the goal of unifying physics, is not a new one. Already the year before Einstein in 1915 introduced his theory of general relativity, Gunnar Nordstrom suggested a unification of gravity and electromagnetism with the introduction of a fifth dimension. These forces were the two only forms of interaction known at that time. But this idea was forgotten for some time with the eruption of the First World War. But in April 1919 Theodor Kaluza introduced independently, in a letter to Einstein, a fifth dimension in an attempt to unify Einstein's theory of gravity and Maxwell's theory of light. Oskar Klein (1926) contributed, in this quest, with his assumption that the extra dimension was compactified. The Kaluza-Klein theory was a fact. This theory includes an extra space dimension that is rolled up into a tiny circle, i.e. compactified. And in this five dimensional theory, there is only one underlying force, gravity. But in the four-dimensional spacetime observed at great distances, it appears to be three kinds of forces, among these a gravitational and an electromagnetic force. This topic was initially a popular topic for research, but lost much of its interest with the introduction of quantum mechanics.

In recent years the topic of extra dimensions has experienced a renewed interest. This renewed interest is also due to the exciting possibility of observing new and spectacular physical phenomena at far lower energy scales than otherwise. Even at energies available in the not so distant future, these phenomena could appear. Among these is the creation of higher dimensional semi-classical microscopic black holes. The possibility of observing these objects, is viewed as an opportunity to perhaps discover new intriguing physics.

Therefore from (i) using Einstein's equivalence principle we get the minimum distance beyond which the force of gravity no longer continues to increase as; $R = \frac{R_s}{2\alpha_g{}^n}$ (ii). Where $R_s = \frac{2GM}{c^2}$ is the Schwarzschild radius.

We therefore conclude that; at n=0 extra spatial dimension, we have a physical theory of General relativity at a length scale of $R = R_s = \frac{GM}{c^2}$ -half of the Schwarzschild radius.

At n=1/2 extra dimension, we have the quantum theory of gravity (New physics) at the Planck length scale $l_p = \sqrt{\frac{\hbar G}{c^3}}$.

At n=1 extra dimension, we have the theory of Quantum mechanics at the Compton wavelength scale of $\lambda = \frac{\hbar}{mc}$.

Lastly at n=2 we have new physics at a length scale $R = \frac{\hbar^2}{GMm^2}$ and the journey continues.

According to the Standard Model of particle physics, the world is governed by four fundamental forces: gravity, electromagnetism, and the weak and strong nuclear forces. Although things act a bit "spooky" down on the quantum level, science has managed to generally describe all of these forces at both the macro and quantum scales – except gravity.

Gravity is the weakest of the fundamental forces, and it's been suggested that this is because some gravitons (the hypothetical particles) that carry the gravitational force tend to escape into extra dimensions. We're simply too big to travel through or even notice these other dimensions.

So, to study whether these extra dimensions are lurking in extremely tiny spaces, the researchers from Osaka, Kyushu and Nagoya Universities set out to test gravity on the sub nanometer scale. To do so, they used the world's highest intensity neutron beam, which is housed at the Japan Proton Accelerator Research Complex (J-PARC).

The team found that the results matched predictions based on the known laws of physics, which indicates that Newton's law still applies as expected down to a scale of less than 0.1 nanometers. No unexplained force ie, another dimension is acting on these particles at this scale.

That doesn't mean those extra dimensions aren't there, just that they may be hiding at even smaller scales still. The researchers are currently working to further improve the sensitivity of the equipment, which might help them probe those tiny spaces.

In a completely different context, an international team of researchers led by Professor Immanuel Bloch (LMU/MPQ) and Professor Oded Zilberberg (ETH Zürich) has now demonstrated a way to observe physical phenomena proposed to exist in higher-dimensional systems in analogous real-world experiments. Using ultracold atoms trapped in a periodically modulated two-dimensional superlattice potential, the scientists could observe a dynamical version of a novel type of quantum Hall effect that is predicted to occur in four-dimensional systems. *(Nature, 4 January 2018)*

"Physically, we don't have a 4D spatial system, but we can access 4D quantum Hall physics using this lower-dimensional system because the higher-dimensional system is coded in the complexity of the structure," a researcher with the US-based team, Mikael Rechtsman from Penn State University, told Ryan F. Mandelbaum at Gizmodo. "Maybe we can come up with new physics in the higher dimension and then design devices that take advantage the higher-dimensional physics in lower dimensions."

The above statements can be summed up in the following simplest model;

Let the Gravitational force between two identical particles be related to the magnetic force between them and similarly let the electric force between two particles be related to the magnetic force as;

$Gravitational\ force\ (\frac{Gm^2}{R^2}) = magnetic\ force\ (Bec) \times \alpha_g{}^n$ (iii) and

$$\text{Electric force } \left(\frac{e^2}{4\pi\varepsilon R^2}\right) = \text{magnetic force (Bec)} \times \alpha_e{}^n$$

(iv)

Where α_e is the electromagnetic coupling constant- Fine structure constant

The magnetic flux, represented by the symbol Φ, threading some contour or loop is defined as the magnetic field **B** multiplied by the loop area, $A=\pi R^2$, i.e. $\Phi = \mathbf{B} \cdot \mathbf{A}$. Obviously, both **B** and **A** can be arbitrary and so is Φ. The inverse of the flux quantum, $1/\Phi_0$, is called the **Josephson constant**, and is denoted K_J.

However, if one deals with the superconducting loop or a hole in a bulk superconductor, it turns out that the magnetic flux threading such a hole/loop is quantized. Therefore the magnetic flux quantum from (iii) and (iv) will be given by $\Phi_G = \pi G m^2/ec\alpha_g{}^n$ and $\Phi_E = e/4\varepsilon c\alpha_e{}^n$ respectively. Such that at n=0 extra dimension, $\Phi_G = \pi G m^2/ec$ and $\Phi_E = e/4\varepsilon c$ representing the classical flux at 3D spatial dimensions.

At n=1/2 extra dimension, $\Phi_G = \frac{\pi m}{e}\left(\frac{G\hbar}{c}\right)^{1/2}$ and $\Phi_E = \left(\frac{\pi \hbar}{4\varepsilon c}\right)^{1/2}$ representing the quantum theory of Gravity.

At n=1 extra dimension, $\Phi_G = \pi\hbar/e$ and $\Phi_E = \pi\hbar/e$ representing the magnetic flux quantum at the quantum scale. Also at n=1 the magnetic flux value is the same in both equations, meaning that the gravitational force becomes analogous to the electromagnetic force at n=1.

In other words, just as a 3D object casts a 2D shadow, scientists have managed to observe a 3D shadow potentially cast by a 4D object – even if we can't actually see the 4D object itself. That could unlock some new findings in the very fundamentals of science.

Dark Matter and the Cosmological constant Problem

How to Calculate a Mysterious Repulsive Force Pulling Galaxies Apart

It is known that in a homogenous cosmological universe, a positive cosmological constant induces repulsive forces. The question is; Is there a classical formula of the force of the cosmological constant like that of the gravitational force? How does the repulsive force relate to the cosmological constant and the coupling constant? How does understanding the energy density in relation to force, change the way we perceive Einstein's field equation? The section sets out to answer these and more questions about the cosmological constant problem.

Dr Lee Smolin represents the perimeter institute for theoretical physics. He claims that the mathematisation of physics has resulted in the reduction of the cosmos to a mathematical entity, which has not only confused physicists but accounts for their worst and most distracting assertions.

There is a wide spread speculation that the mathematical formulation of physics has not only confused physicists but has also lead to failures in the development of a quantum theory of gravity.

Although both general relativity and quantum mechanics work well in the domain of their applicability, it's unfortunate that there is no unified theory of gravity with quantum mechanics.

It is proposed that the unification of gravity with quantum mechanics will require us to change the kind of mathematics that was used by either Einstein, Schrödinger and Hawking in the development of both theories. But why do we bother at all if there is another way in which we can express the theory better without the use of tensor fields.

The problem with the mathematical formulation of general relativity if at all it exists stems from the non existence of its experimental observation which wasn't the case with quantum mechanics. The formulation of quantum mechanics was based on the existence of experimental observations. Therefore quantum mechanics was founded on the existence of experiments which wasn't the case with general relativity. Einstein had to base his theorization on thought experiments which could or wouldn't be nearer to any experimental confirmation of the phenomenon being studied.

The same is also true for the formulation of quantum gravity. There is no sound experimental proof for the existence of quantum gravitational effects and therefore scientists like Hawking Stephen have also clung to the old formulations that were used by Einstein and his contemporaries to develop a quantum theory of gravity.

In this brief section we show that an existence of a unified theory is rooted deep into the unnoticed pressure-energy density similar to the stress energy tensor appearing in Einstein field theory.

Our major aim therefore is to provide proof for the questions set out below;

If the cosmological constant introduces a force of repulsion between bodies. Is it true that the force increases in simple proportion to the cosmological constant and the coupling constant.

Is there a classical formula of the force of the cosmological constant like that of the gravitational force?

Einstein's general relativity equations famously described the curvature of space-time as the mechanism for gravity. In the original theory, Einstein added a "cosmological constant" that acted as an expulsive force to counteract gravity. That stabilized the universe so it didn't collapse in on itself, but Einstein abandoned the idea when further astronomical observations showed the universe was accelerating and not static, as the great physicist had thought.

Analogous to the known Einstein field equation, the curvature of space Λ (cosmological constant) is here related to the energy density ω as,

$$\Lambda = \kappa\omega = \kappa \left(\frac{F^2}{8\pi\alpha\,hc} \right)$$

Where $\kappa = \frac{8\pi G}{c^4}$ is a constant appearing in Einstein's field equation, F is the force in an interaction and α is the coupling constant.

The above expression implies that the cosmological constant is related to the force and therefore increases as a square of the force.

For the energy density in electric field, where $F = Ee$ and $\alpha = \frac{e^2}{4\pi\varepsilon\hbar c}$, the energy density will be given by, $\omega = \frac{F^2}{8\pi\alpha\hbar c} = \frac{\varepsilon E^2}{2}$.

While for the energy density in the gravitational field, where $F = mg$ and $\alpha = \frac{Gm^2}{\hbar c}$, the energy density will be given by, $\omega = \frac{F^2}{8\pi n\hbar c} = \frac{g^2}{8\pi G}$. This can be written in simple terms as $\omega = \frac{ng^2}{2}$ where $\eta = \frac{1}{4\pi G}$.

From (1) therefore, the force responsible for the expansion of the universe is related to the cosmological constant by,

$$F = E_{pl}(\alpha\Lambda)^{1/2}$$

Where $E_{pl} = 1.9605 \times 10^9 J$ is the Planck energy.

Given the Planck (2015) values of $\Omega_\Lambda = 0.6911 \pm 0.0062$ and $H_o = 67.74 \pm 0.46$ (km/s)/Mpc $= (2.195 \pm 0.015) \times 10^{-18}$ s^{-1}, Λ has the value of $1.11 \times 10^{-52} m^{-2}$ as given in wikimedia commons.

Based on the above given value, the force will then have a value of

$$F_{ob} = 2.0655 \times 10^{-17}(\alpha)^{1/2}$$
$$F_{ob} \sim 1.8 \times 10^{-18} N$$

This therefore is a force responsible for the expansion of the Universe. It is such a small force that will require sophiscated machines to measure. While the above force value is based on the fine structure constant, there is a value that is even smaller than that value by,

$F_{ob} \sim 1.58 \times 10^{-36} N$ at $\alpha = 5.87 \times 10^{-39}$ between two protons.

However in quantum electrodynamics (QED) we compute a much larger value of

$F_{QED} \sim 2.82 \times 10^{44} (\alpha)^{1/2} N.$

This huge discrepancy is known as the cosmological constant problem. Therefore the relative strength of the force will be given by;
$$\frac{F_{QED}}{F_{ob}} \sim 10^{61}$$
The above value is in agreement with the Hubble age to the Planck time, which is the same as the total mass of the universe to the Planck mass as,
$$\frac{F_{QED}}{F_{ob}} = \frac{t_H}{t_{pl}} = \frac{M_U}{M_{pl}} \sim 10^{61}$$
The above given relationship implies a persistence constant error that is evident when comparing observational and theoretical calculations. This error needs to be distributed uniformly in order to correct for large discrepancies which accrue to calculated values in relation to observed values.

The problem lies in knowing the observed force value to the calculated value, since the force ratio doesn't correspond to the other ratios of time and mass. In other words changing the ratio $\frac{F_{QED}}{F_{ob}}$ to $\frac{F_{ob}}{F_{QED}}$ will cause other ratios to change.

It is therefore observed that the ratio of Hubble age to the Planck time and the total mass of the universe to the Planck mass will only be in line or tally with the Planck force to the Hubble force by a value $\sim 10^{61}$ and not otherwise.

Keeping other factors constant it is clear from the above given observations that the mysterious, repulsive force pulling galaxies apart is proportional to the coupling constant value in a given interaction. This proposal will be of such a great importance to the work of researchers involved in the field of quantum gravity.

Why does the Zero-Point Energy Of the Vacuum not cause a Large Cosmological Constant? What Cancels it out?

The cosmological constant problem is the large discrepancy between the experimental observed value of the cosmological constant and its theoretical calculated value.

Whereas observations give values of the energy density and the cosmological constant as,

$\omega \approx 1.18 \times 10^{-9} J/m^3$, $\Lambda \approx 1.11 \times 10^{-52} m^{-2}$ (a)

The calculated theoretical values of the energy density and the cosmological constant are larger by,

$\omega \approx 2.531 \times 10^{114} J/m^3$, $\Lambda \approx 5.23 \times 10^{71} m^{-2}$ (b)

The question is; why does the Zero-point energy of the vacuum not cause a large Cosmological constant? What cancels it out?

Before we proceed, let me show you how the above given values for the energy density and cosmological constant come about,

Consider an expanding universe, where a test particle of mass m_i in this universe subject to a force F is accelerating with a constant acceleration a. The energy density ω in this case is related to the force by,

$$\omega = \frac{F^2}{8\pi\alpha\hbar c}$$

Where, $= m_i a$, α is the coupling constant

This gives,
$$\omega = \frac{m_i^2 a^2}{8\pi\alpha\hbar c} \quad (1)$$

Also, the cosmological constant is related to the acceleration by,
$$\Lambda = \frac{a^2}{\alpha c^4} \quad (2)$$

To obtain the experimental observed values of energy density and the cosmological constant, we assume the following values of mass and acceleration

$m_i = 2.18 \times 10^{-8} kg$, Planck mass
$a = 1.2 \times 10^{-10} m/s^2$, small acceleration (MOND by Milgrom)
$\alpha = 1/137$, Fine structure constant

Substituting these values in (1) and (2) we surely get values in (a)

To obtain the theoretical values of energy density and the cosmological constant, we assume the following values of mass and acceleration

$m_i = 2.18 \times 10^{-8} kg$, Planck mass
$a = 5.5608 \times 10^{51} m/s^2$, maximal acceleration (Planck units)
$\alpha = 1/137$, Fine structure constant

Substituting these values in (1) and (2) we surely get values in (b)

We observe that, the common factor in the above calculations is the Planck mass and the coupling constant. This implies that, to solve the cosmological constant problem we need to create a relationship between the energy density and the cosmological constant where the acceleration and coupling constant disappears as,

$$\Lambda = \frac{8\pi h}{m_i^2 c^3}\omega \quad (3)$$

$$\Lambda = \frac{9.82 \times 10^{-59}}{m_i^2}\omega$$

This equation means that, the zero point energy of the vacuum will not cause a large cosmological constant because of the inverse square law given above. This means that, the large mass of the interacting particle cancels out the large cosmological constant caused by the zero point energy density of the vacuum.

For a small interacting particle $m_i \approx 10^{-70} kg$, we get a large cosmological constant and vice versa. This means that, all observations for the cosmological constant assume the interacting particle to be the Planck mass. Implying that the Planck mass is the least mass responsible for the acceleration of the universe, below the Planck mass, the universe becomes static. Therefore the above equation can be used to determine the mass of a particle responsible for the expansion of the universe by assuming a constant cosmological constant.

Proof of Newton's Law of Universal Gravitation

Previously we showed that the energy density ω is related to the force F by,

$$\omega = \frac{F^2}{8\pi\alpha\hbar c}$$

Where, $\hbar$ is the reduced Planck constant, c is the constant speed of light and is the coupling constant or principal quantum number

In this short notice we clearly prove that F is the gravitational force that was put forward by Newton as we are yet to see below,

Independent of the mass and distance, the force between two particles as in the case of the Casmir affect is therefore given by,

$$F = \sqrt{8\pi\alpha\hbar c\omega} \qquad (1)$$

Below I give two conditions on which the above force will reduce to the Newton's universal law of gravitation

The energy density is related to the cosmological constant by,

$$\omega = \frac{m^2 c^3}{8\pi \hbar} \Lambda$$

This was previously derived and m was the Planck mass

Here m is taken to represent the mass of the particle in circular orbit around a massive body of mass M at a distance or radius of curvature R from M. Where, $\Lambda = \frac{1}{R^2}$.

The coupling constant or principal quantum number is here given as a ratio of the areas as,

$$\alpha = \frac{A_s}{A}$$

Where $A_s = \frac{4\pi G^2 M^2}{c^4}$, is the Schwarzschild area occupied by a massive body and $A = 4\pi R^2$ is the total area of circular orbit of a mass m.

Substituting condition 1 and 2 into equation (1) above we obtain the Newton's law of gravitation as,

$$F = \frac{GMm}{R^2}$$

This derivation is proof that gravity is a result of the quantum vacuum energy density. While this has proved to be a short insight into the emergency of gravity, we shall have a long discussion of this research in the coming chapters.

Revised Gravitation Theory for the Modified Newtonian Dynamics (MOND) Paradigm and Beyond

To differ from the 1982 Milgrom Hypothesis we introduce a new force law as an alternative to Newton's law of gravitation and Einstein's general relativity. Using this new force law, we deduce Hawking Black hole temperature which can't be proved under the Milgrom hypothesis. We also recover the Tully – Fisher relationship from which the angular frequency and energy of the graviton- photon oscillations are determined. This new force law is important because it recovers the laws of a black hole, something that has failed the other MOND alternatives.

It must be noted that the Milgrom Hypothesis that was introduced in 1982 to account for the flat rotation curves of spiral galaxies under the assumption of small acceleration fails to deduce the temperature and entropy of a black hole. In my opinion I think that all alternatives to Newton's gravity or General relativity should at least be able to explain details near or beyond black hole singularities if at all they exist. Failure for MOND to deduce the Hawking laws for black holes is one of those major indicators that it is not a genuine alternative to GR. Even if Black holes haven't been observed, I think it is in order for MOND to be consistent with the other laws of physics.

The Modified Force Law

The New force law is

$$F = \frac{me}{R}\sqrt{\frac{GM\omega}{4\pi\hbar\varepsilon_o}} \qquad (1)$$

Where $\omega=2\pi f$ is the angular frequency of the graviton-photon oscillations and e is the charge on an electron. The above given law was used in chapter 10 and chapter 11 of the book "Quantum Gravity in a Nutshell1"

In a limit of $\omega = \frac{GM}{R^2}\left(\frac{4\pi\hbar\varepsilon_o}{e^2}\right) = \frac{g_N}{c\alpha_e}$, where g_N is the usual Newtonian acceleration due to gravity and α_e is the fine structure constant, the above new force law reduces to the Newtonian law of universal gravitation.

The Tully-Fisher Relation

One of the best fit predictions of MOND is a single universal Tully-Fisher relation.
" The relation between asymptotic velocity and the mass of the galaxy is an absolute one" (Milgrom 1983). This is given by, $V^4 = a_o GM$, where $a_o = 1.2 \times 10^{-10} ms^{-2}$. In this paper an equation similar to the Tully-Fisher relation is deduced from (1) as given below,

For circular orbits about a point mass, M

$$\frac{V^2}{R} = \frac{e}{R}\sqrt{\frac{GM\omega}{4\pi\hbar\varepsilon_o}}$$

This gives an asymptotically rotation velocity independent of R:

$$V^4 = \left(\frac{e^2\omega}{4\pi\hbar\varepsilon_o}\right)GM = \omega c\alpha_e GM \qquad (2)$$

It is this behavior that gives rise to asymptotically flat rotation curves and the Tully-Fisher relation (Tully & Fisher 1977)

Comparing (2) to the Tully-Fisher relation, we determine the angular frequency of the graviton oscillation as,

$$\omega = \frac{a_o}{c\alpha_e} = 5.48 \times 10^{-17}\,rad/s$$

Graviton-Photon energy

It has been known that the energy of a photon is related to the angular frequency by, $E = \hbar\omega$. This equation worked well at the atomic scale but it could not explain the spectrum of the galaxy clusters. Here we give a formula that works at the galactic scale as given below,

From (2) $\omega = \frac{V^4}{c\alpha_e GM}$, then

$$E = \hbar\omega = \frac{\hbar V^4}{c\alpha_e GM} \qquad (3)$$

Comparing this to the total energy of the gravitating body from Einstein Energy relation, we have

$$Mc^2 = \frac{\hbar V^4}{c\alpha_e GM}$$

This then gives,

$$\frac{V^4}{c^4} = \alpha_e \alpha_g$$

Where, $\alpha_g = \dfrac{GM^2}{\hbar c}$ is the gravitational coupling constant

This shows that the velocity of the stars in circular orbit about a center of mass M will never exceed that of light by, $V = c(\alpha_e \alpha_g)^{1/4}$.

Derivation of the Temperature of a Black Hole from the new force law

A Black hole evaporates this way; the gravitational force of the entire mass of a black hole balances the electromagnetic forces between individual electrons inside the Black hole. When this happens, electrons inside a black hole acquire the same charge sign as with those at the surface and therefore a particle is emitted from the surfaces of a black hole

$$\dfrac{e^2}{4\pi \hbar \varepsilon_o R^2} = \dfrac{me}{R} \sqrt{\dfrac{GM\omega}{4\pi \hbar \varepsilon_o}}$$

Squaring both sides of the above equation and arranging, in a limit v=c, we have the kinetic energy of the emitted particle as,

$$KE = mc^2 = \dfrac{4e^2 \lambda \mu_o \hbar c^3}{8\pi GmMA}$$

Where λ is the wavelength of a particle emitted from a black hole, μ_o is the permeability of free space and

$A = 4\pi R^2$ is the surface area of the event Horizon of a Black hole.

By equivalence, $kT = \dfrac{4e^2 \lambda \mathbb{Z}_o \hbar c^3}{8\pi GmMA}$

Where kT is the thermal kinetic energy of the emitted particle, then the temperature of the emitted particle is,

$$T = \dfrac{4e^2 \lambda \mathbb{Z}_o}{mA} \left(\dfrac{\hbar c^3}{8\pi GMk} \right)$$

In a limit, $= \dfrac{mA}{4e^2 \mathbb{Z}_o}$, the above equation gives the temperature of a black hole as,

$$T = \left(\dfrac{\hbar c^3}{8\pi GMk} \right)$$

In conclusion, the Milogram Hypothesis that works well on a classical level fails to work on a quatum level. In this kind of situation a theory as given above is required to explain details where the quantum gravitation effects become important. This still points to the requirement of a quantum theory of gravity. In this paper, we have deduced a near approximation to the quantum theory of gravity which works well in explaining the spiral galaxy rotation curves and the energy associated with the quanta emitted in the process.

Planck Stars

Evidence for Maximal acceleration and Singularity Resolution in Quantum Gravity

The appearance of singularities in any physical theory is an indication that either something is wrong or we need to reformulate the theory itself. Singularities are like dividing something by zero. One such theory plagued by singularities is the General theory of relativity (GR) and the problems in GR arise from trying to deal with a universe that is zero in size (infinite densities). However, quantum mechanics suggests that there may be no such thing in nature as a point in space-time, implying that space-time is always smeared out, occupying some minimum region. The minimum smeared-out volume of space-time is a profound property in any quantized theory of gravity and such an outcome lies in a widespread expectation that singularities will be resolved in a quantum theory of gravity. This implies that the study of singularities acts as a testing ground for quantum gravity.

Loop quantum gravity (LQG) suggests that singularities may not exist. LQG states that due to quantum gravity effects, there must be a minimum distance beyond which the force of gravity no longer continues to increase as the distance between the masses become shorter or alternatively that interpenetrating particle waves mask gravitational effects that would be felt at a distance. It must also be true that under the assumption of a corrected dynamical equation of LQ cosmology and brane world model, for the gravitational collapse of a perfect fluid sphere in the commoving frame, the sphere does not collapse to a singularity but instead pulsates between a maximum and minimum size, avoiding the singularity.

The resolution of classical singularities under the assumption of a maximal acceleration has been studied using canonical methods for Rindler, Schwarzschild, Reissner-Nordstrom, Kerr-Newman and Friedman-Lemaitre metrics.

The demand for consistency between a quantum description of matter and a geometric description of spacetime, as well as the appearance of singularities (where curvature length scales become microscopic), indicate the need for a full theory of quantum gravity. For example; for a full description of the interior of black holes, and of the very early universe, a theory is required in which gravity and the associated geometry of space-time are described in the language of quantum physics. Despite major efforts, no complete and consistent theory of quantum gravity is currently known, even though a number of promising candidates exist.

The first step towards the development of a quantum theory of gravity lies in studying the kind of physics behind white dwarfs, neutron stars and black holes which are born when normal stars die. White dwarfs are supported by the pressure of degenerate electrons, Neutron stars are supported largely by the pressure of degenerate neutrons while Black holes on the other hand, are completely collapsed stars that is, stars that could not find any means to hold back the inward pull of gravity and therefore collapse to a singularity.

This section is aimed at answering questions like; i) Do objects continually collapse to a singularity or there is a limiting distance below which the very notions of space and length cease to exist?

Theorem:- A star more than three times the size of our Sun collapses in this way; the gravitational forces of the entire mass of a star overcomes the electromagnetic forces of individual atoms and so collapse inwards. If a star is massive enough it will continue to collapse creating a Black hole, where the whopping of space time is so great that nothing can escape not even light, it gets smaller and smaller. The star in fact gets denser as atoms even subatomic particles literally get crashed into smaller and smaller space, and it's ending point is of course a space time singularity.

One of the first and crucial step towards the development of a quantum theory of gravity is the resolution of singularities that plague the Einstein General theory of Relativity. The solution has been given in part by Carlo Rovelli and Francis Viddoto in LQG but in this book we develop a different approach towards the problem as;

Let the Heisnberg Uncertainity principle be modified and written as, $r \times p = \frac{\hbar}{2}\alpha^n$ Where r represents the size of a star, in this case-horizon radius, p is the momentum of a particle approaching or falling into the hole of a star, α is the coupling constant and n is positive.

From the modified uncertainty relation given above, we prove the existence of a maximal acceleration which in turn yields a bound on temperature, curvature and on the energy density in appropriate cosmological contexts, supporting the results in LQ Cosmology and for Black holes.

Maximal acceleration

From the uncertainity relation we deduce the acceleration as,

$$a = \frac{c^2}{2r}\alpha^n$$

Taking, $\alpha = \frac{GM^2}{\hbar c} = \left(\frac{M}{M_{pl}}\right)^2$ (Gravitational coupling constant, where M_{pl} is Planck mass) and $r = \frac{2GM}{c^2}$ for a Schwarzichild Black hole, the maximal acceleration that a particle can have when **n=1/2** will be given by,

$$a = \left(\frac{c^7}{16G\hbar}\right)^{1/2}$$

While for $r = \left(\frac{Ge^2}{4\pi\varepsilon c^4}\right)^{1/2}$ for a Reissner Nordstrom Black hole. Where e is charge on an electron, ε is the permittivity of free space and $\hbar$ is the reduced Planck constant. We have a maximum acceleration of $a_{max} = \frac{M}{e}\left(\frac{\pi\varepsilon c^7}{\hbar}\right)^{1/2}$.

These acceleration also imposes a general upper bound on temperature given by; From the Unruh temperature $T = \frac{\hbar a}{2\pi c k}$, If we subsititute for the maximal acceleration we have, $T = \frac{1}{8\pi k}\left(\frac{c^5\hbar}{G}\right)^{1/2}$ for a Schwarzschild Black hole and $T = \frac{M}{ke}\left(\frac{\hbar\varepsilon c^5}{4\pi}\right)^{1/2}$.

Therefore a maximal acceleration screens the classical singularity.

Cosmological Energy Density

To resolve the singularities of General relativity we consider the possibility that the energy of a collapsing star and any additional energy falling into the hole could condense into a highly compressed core with density of the order of the Planck density. If this is the case, the gravitational collapse of a star does not lead to a singularity but to one additional phase in the life of a star: a quantum gravitational phase where the gravitational attraction is balanced by a quantum pressure.

Since the energy density is expressed as force per unit surface area of a star. Where the force is mass times acceleration we have, $\rho = \frac{ma}{4\pi R^2}$ which gives, $\rho = \frac{mc^2}{8\pi R^3}\alpha^n$ and on conditions that $R = \frac{2GM}{c^2}$, $\alpha = \frac{GM^2}{\hbar c} = \left(\frac{M}{M_{pl}}\right)^2$ and n=1 we have a maximum energy density value as, $\rho = \frac{c^7}{\pi\hbar(8G)^2}$. Nature appears to enter the quantum gravity regime when the energy density of matter reaches the Planck scale. The point is that this may happen well before relevant lengths become planckian. For instance, a collapsing spatially compact universe bounces back into an expanding one.

The bounce is due to a quantum gravitational repulsion which originates from the Heisenberg uncertainty, and is akin to the "force" that keeps an electron from falling into the nucleus. The bounce does not happen when the universe is of planckian size, as was previously expected; it happens when the matter energy density reaches the Planck density as derived above. For example, if we let the acceleration due to gravity of a star be in eqillibrium with the acceleration of the quantum particle falling into a hole (From modified UP), then the size of a star will be given following the derivation below,

$$\frac{GM}{R^2} = \frac{c^2}{2r} \alpha^n$$

Where R is the distance between a star and a particle and r is the size of a star (From the uncertainty principle)
Then,

$$r = \frac{c^2 R^2}{2GM} \alpha^n = \frac{R^2}{R_s} \alpha^n$$

Where $R_s = \frac{2GM}{c^2}$ is the radius of the event Horizon of a star.

Therefore on condition that, $\frac{R^2}{R_s} = l_p$ (Planck Length) and $\alpha = \left(\frac{M}{M_{pl}}\right)^2$ we have the size of a star as, $r = \left(\frac{M}{M_{pl}}\right)^{2n} l_p$. Where M is the mass of the star and n is positive. For instance, if n = 1/6, a stellar-mass black hole would collapse to a Planck star with a size of the order of 10^{-10} centimeters. This is very small compared to the original star in fact, smaller than the atomic scale but it is still more than 30 orders of magnitude larger than the Planck length. This is the scale on which we are focusing here. The main hypothesis here is that a star so compressed would not

satisfy the classical Einstein equations anymore, even if huge compared to the Planck scale. Because its energy density is already planckian(for more examples see Carlo. R & F. Vidotto 2014-Planck Stars).

In conclusion the relation $\frac{R^2}{R_s} \geq l_p$ as has been given above states that lengths beyond the Planck length are meaningless and that the singularities are resolved at the Planck scale, which in this case is another form of the uncertainty principle. The results given above therefore indicate the requirement for the modification of the Uncertainty principle.

References

E.Benedetto, A.Feoli (2015) Unhru temperature with maximal acceleration. Modern Physics letters A. DOI: 10.1142/S0217732315500753

Conor Purcell (2018) Quantum gravity and the small matter of a theory of everything. The Irish Times, www.irishtimes.com

https://en.wikipedia.org/wiki/Wikipedia:Text_of_Creative_Commons_Attribution-ShareAlike_3.0_Unported_License

O. Aursjo (2005). microscopic Black holes and extra dimensions

C. Gao and Y.Lu, Pulsations of a black hole in LQG (2012) arXiv:1706.08009v3

A.H. Chamseddine and V.Mukhanov, Non singular black hole (2016) arXiv 1612.05861v1
M.Bojowald and G.M.Paily, A no-singularity scenario in LQG (2012) arXiv: 1206.5765v1

P.Singh, class.Quant.Grav,26,125005(2009), arXiv:0901.2750

P.Singh and F.Vidotto, Phys.Rev, D83,064027(2011) arXiv:1012.1307

C.Rovelli and F.Vidotto, Phy. Rev,111(9) 091303(2013) arXiv:1307.3228v2

M.Bojowald, Initial conditions for a universe, Gravity Research Foundation (2003)

A.Ashtekar, Singularity Resolution in Loop Quantum Cosmology (2008) arXiv:0812.4703v1

J.Brunneumann and T.Thiemann, On singularity avoidance in Loop Quantum Gravity (2005) arXiv:0505032v1

L.Modesto, Disappearence of the Black hole singularity in Quantum gravity (2004) arXiv:0407097v2

Thank you very much for buying my book.

You have supported my research works towards the unification of gravity with quantum mechanics.

www.ingramcontent.com/pod-product-compliance
Lightning Source LLC
Chambersburg PA
CBHW030518220526
45464CB00006B/2855